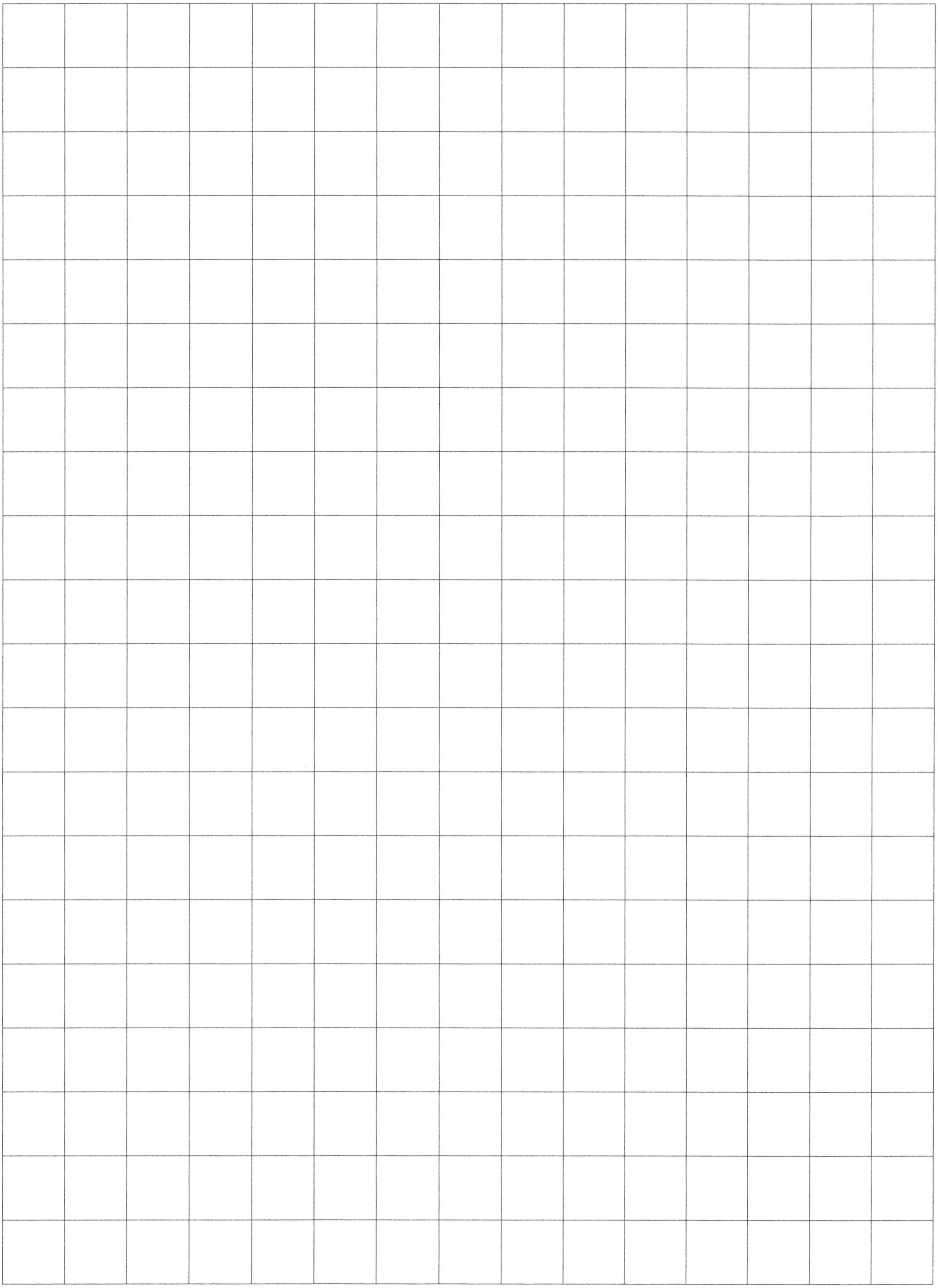

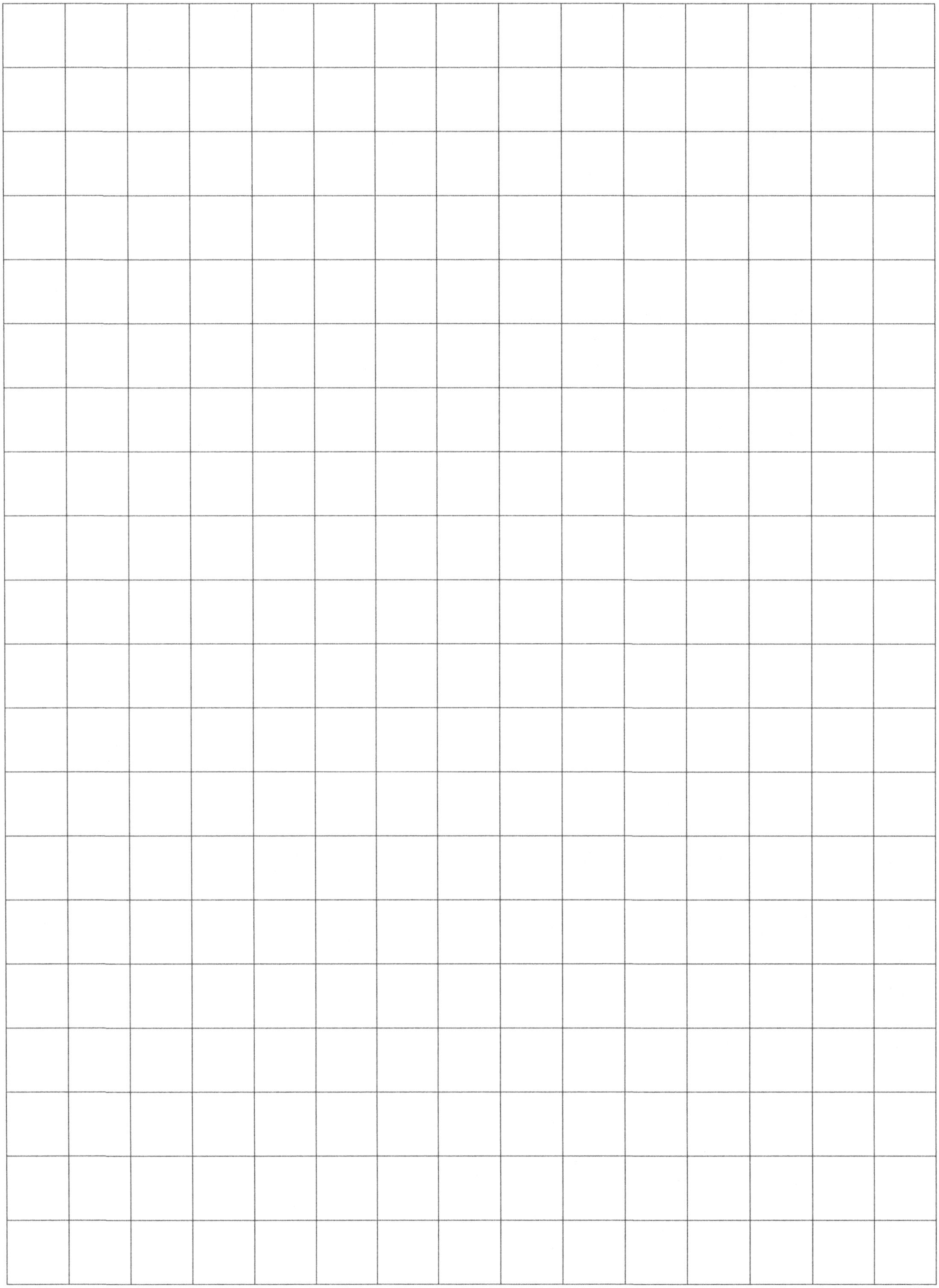

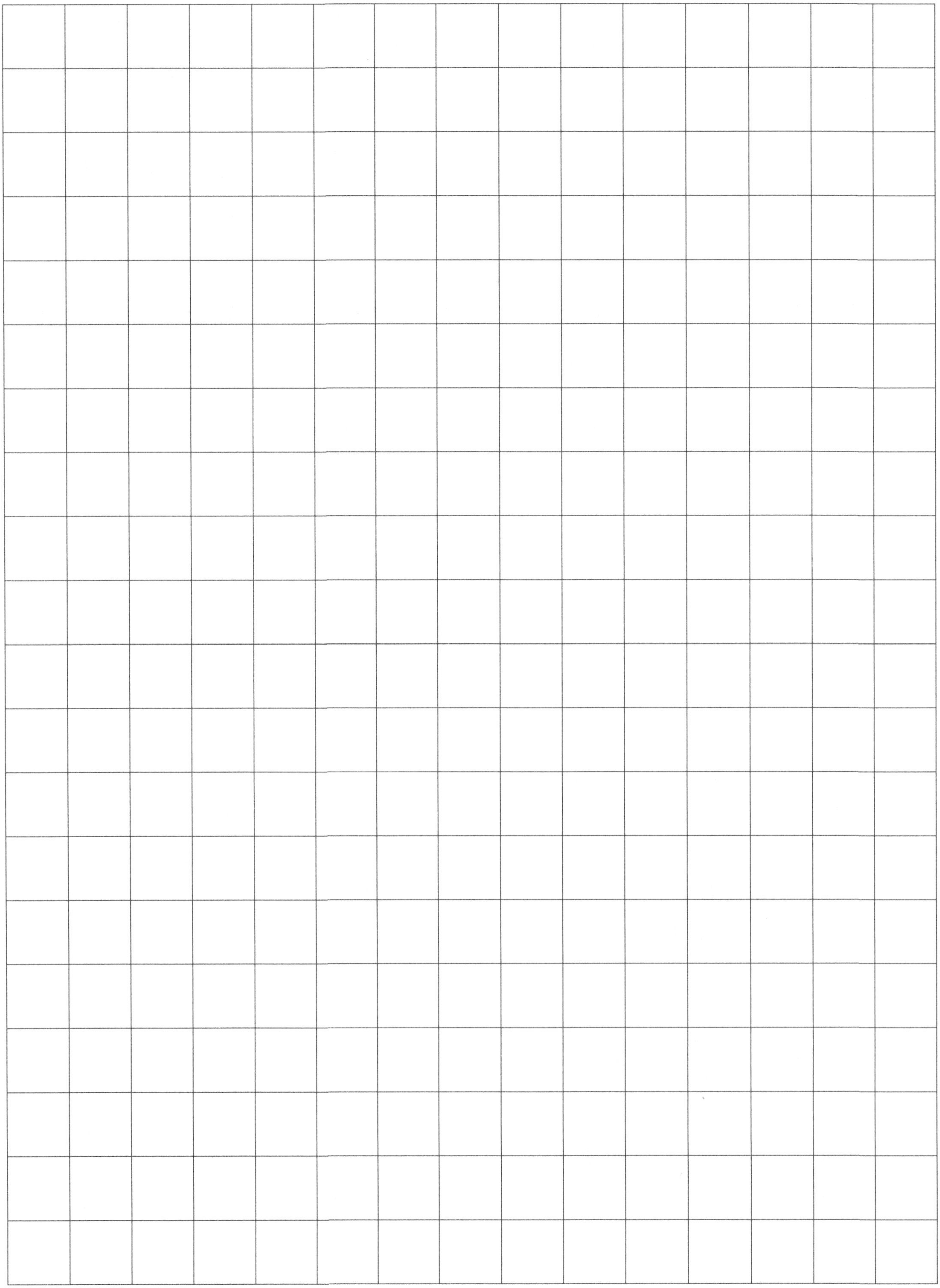

Made in the USA
Columbia, SC
15 June 2023

18080991R10057